LEVEL
3
Fact Reader

Bling!

100 FUN Facts About Rocks and Gems

Emma Carlson Berne

NATIONAL GEOGRAPHIC

Washington, D.C.

To my mom, who loves rocks, minerals, all parts of the natural world —E.C.B.

Published by National Geographic Partners, LLC, Washington, DC 20036

Designed by Yay! Design

The author and publisher gratefully acknowledge the expert content review of this book by John Cottle, Ph.D., professor of geology, University of California, Santa Barbara, and the literacy review of this book by Mariam Jean Dreher, professor emerita of reading education, University of Maryland, College Park.

Photo Credits

AS: Adobe Stock; GI: Getty Images; NGIC: National Geographic Image Collection; SS: Shutterstock
Cover (main), Adrienne Bresnahan/GI; (LO), J-Palys/GI; heading (throughout), Julie Boro/AS; 1, LVV/SS; 3, Sebastian Janicki/SS; 4 (UP LE), Denis Comeau/SS; 4 (UP RT), Tom Reichner/SS; 4 (CTR), Mega Pixel/SS; 4 (LO LE), John Cancalosi/Alamy Stock Photo; 4 (LO RT), Marisabell/Can Stock Photo Inc.; 5 (UP LE), CE Photography/SS; 5 (UP CTR), Alhovik/Dreamstime; 5 (UP RT), Museum of East Asian Art/Heritage Images/GI; 5 (CTR LE), INTERFOTO/Alamy Stock Photo; 5 (CTR RT), DougVonGausig/GI; 5 (LO LE), Nik_MerkulovAS; 5 (LO RT), Mushika/GI; 6, Coldmoon_photo/GI; 7 (UP), Minakryn Ruslan/AS; 7 (LO), Tryfonov/AS; 8, Historia/SS; 9 (UP), Reid Dalland/AS; 9 (LO), jonnysek/AS; 10-11, camera-withlegs/AS; 11 (UP), Dan Breckwoldt/AS; 11 (LO), Nicholas/AS; 12, David Steele/SS; 13 (UP), Ahmad Faizal Yahya/SS; 13 (LO), Ozbalci/GI; 14, soft_light/AS; 15, ttsz/GI; 16 (UP), vvoe/SS; 16 (LO), Adam-bowers/Dreamstime; 17 (UP), MNStudio/AS; 17 (LO), sandatlas/SS; 18, oversnap/GI; 19 (LE), Merlin74/SS; 19 (RT), Horst Mahr/GI; 20-21, Matauw/AS; 22, Carsten Peter/Speleoresearch & Films/NGIC; 23, PHOTO RF/Science Source; 24 (UP LE), carlosdela-calle/SS; 24 (UP RT), hsvrs/GI; 24 (LO), mermaidb/iStockphoto; 25 (UP), Breck P. Kent/SS; 25 (LO), KarSol/AS; 26 (UP LE), vvoe/AS; 26 (UP RT), andy koehler/AS; 26 (LO), Lebazele/GI; 27 (UP LE), Pesh Siri/AS; 27 (UP RT), santanor/GI; 27 (LO), Ailani Graphics/SS; 28, medaacek/iStockphoto; 29 (UP), Joel Arem/Science Source; 29 (LO), Manamana/SS; 30-31, Terryfic3D/GI; 31, Martin/AS; 32, Design Pics Inc/NGIC; 33, DEA/G. Cigolini/De Agostini via GI; 34, Eduardo F Guevara/AS; 35 (UP), Monty Rakusen/GI; 35 (LO), Dakota Matrix; 36, Jun Zhang/GI; 37 (UP), nattul/GI; 37 (CTR), Cindy Miller Hopkins/Danita Delimont/Alamy Stock Photo; 37 (LO), alter_photo/AS; 38, Dr. Robert Lavinsky/The Arkenstone; 39 (UP), Olivier Goujon/Alamy Stock Photo; 39 (LO), foxdan-nit/AS; 40, Tom Hanley/Alamy Stock Photo; 41 (UP), Remy De La Mauviniere/AP/SS; 41 (LO), SS; 42, Rogers Fund, 1908/Metropolitan Museum of Art; 43 (LE), Prisma Archivo/Alamy Stock Photo; 43 (RT), Gilbert Carrasquillo/FilmMagic/GI; 44 (UP LE), The Art Archive/SS; 44 (UP RT), marcel/AS; 44 (LO LE), udra/SS; 44 (LO RT), hjschneider/AS; 45 (UP), Jag_cz/SS; 45 (CTR LE), Andrey Kuzmin/AS; 45 (CTR RT), DiamondGalaxy/AS; 45 (LO LE), Garry DeLong/AS; 45 (LO RT), smiltena/AS

Library of Congress Cataloging-in-Publication Data

Names: Berne, Emma Carlson, author.
Title: Bling! : 100 fun facts about rocks and gems / Emma Carlson Berne.
Description: Washington, D.C. : National Geographic Kids, 2022. | Series: National geographic readers | Includes index. | Audience: Ages 6-9 | Audience: Grades 2-3
Identifiers: LCCN 2019055270 (print) | LCCN 2019055271 (ebook) | ISBN 9781426338908 (paperback) | ISBN 9781426338915 (library binding) | ISBN 9781426338922 (ebook)
Subjects: LCSH: Rocks--Juvenile literature. | Gems--Juvenile literature.
Classification: LCC QE432.2 .B48 2022 (print) | LCC QE432.2 (ebook) | DDC 552--dc23
LC record available at https://lccn.loc.gov/2019055270
LC ebook record available at https://lccn.loc.gov/2019055271

Printed in the United States of America
21/WOR/1

Contents

1 A piece of rock has to be at least .16 inch wide in order to be called a pebble.

2 The mineral cobalt got its name from the German word *kobold*, a mischievous goblin said to live in mines.

3 Archaeologists in Texas, U.S.A., found a 1,500-year-old piece of fossilized human poop that contained the bones of a rattlesnake.

4 Powdered aluminum is a main ingredient in rocket fuel.

5 When sand is mixed with other ingredients and heated above 2400 degrees Fahrenheit, it melts and turns into glass.

6 Some people scrub and clean their skin with the volcanic rock pumice.

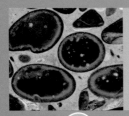

7 British settlers in Ontario, Canada, called one rock "puddingstone" because it reminded them of a dessert.

8 Diamond is the hardest natural material on the planet.

9 Soapstone gets its name from its greasy, or "soapy," feel.

10 The mineral cinnabar has sometimes been called "dragon's blood" because of its deep red color.

25 COOL FACTS ABOUT ROCKS AND

11 When lightning strikes sand, the sand sometimes melts, forming hollow, glass-lined tubes called fulgurites (FUL-gyer-ites).

12 Oil can be extracted from the rock shale.

13 At the 2019 Oscars, pop star Lady Gaga wore a $30 million yellow diamond necklace.

14 The valuable gold in Fort Knox was transported there in 1937 on a heavily guarded nine-car train.

15 Stardust containing minerals rich in iron and nickel regularly falls to Earth.

16 Freezing weather can break apart rocks.

17 Prehistoric peoples made axe-heads from the mineral jade to use at special ceremonies.

18 The medieval healer Hildegard von Bingen believed that licking a sapphire would make a person smarter.

19 An average American uses about 40,000 pounds of minerals every year, including 338 pounds of salt and 28 pounds of aluminum.

20 Not all rocks are hard—clay is soft because it's made up of tiny bits of rock clinging together!

21 Ancient Egyptians said the gem tourmaline passed through a rainbow, giving it the many colors it comes in.

22 Volcanic rocks like trachyte (TRAK-ite) are so hard that the ancient Romans built roads with them.

23 Mineral crystals can look like ferns or like the spokes on a bicycle.

24 The mineral mica gives some makeup its shimmer.

25 The mineral ulexite (YOO-lek-site) is called "television stone"—images or words placed underneath the clear mineral seem to appear on its surface.

Ulexite

GEMS

CHAPTER 1

Rocking Out!

Rocks and minerals can be found throughout our world—and EVEN IN SPACE!

Our planet contains OVER 5,000 known minerals.

the mineral bismuth

Rocks and minerals bubble up from volcanoes. They sit on the ocean floor. They even float through the air! But what are they, exactly?

↑ Some metals—such as copper, silver, and gold— are minerals, too. They can exist on their own or be mixed with other minerals.

Minerals are solids that form in nature. Each mineral is made up of its own mix of chemical elements. And it always forms in an exact crystal structure.

Rocks are made of a mixture of minerals. They can also contain bits of other rocks or fossilized plants or animals. Rocks can be as tiny as a grain of sand or as huge as a boulder.

Rocks from other planets sometimes SLAM INTO EARTH in the form of meteorites.

People started using rocks to make tools as far back as TWO MILLION YEARS AGO.

During the Stone Age, people made axes out of flint. But early humans also used rocks and minerals in other ways. Ancient Egyptians used the green mineral malachite as eye makeup. And in ancient Assyria and Babylonia, rocks like lapis lazuli were thought to protect against spirits that caused illness.

8

The Navajo people in North America have been mining and using turquoise since AT LEAST 200 B.C.

Rocks and minerals are everywhere in our lives today, too. Slate is used for roofing. Talc makes the powdery coating on chewing gum. Humans even have to *eat* a mineral—salt—to survive!

salt crystal

The mineral silver is used in MEDICAL BANDAGES to prevent infection.

Some of the giant rocks of Stonehenge weigh about 50 tons each—that's AS MUCH AS 12 HIPPOS!

Prehistoric people built enormous rock structures that still stand today. For centuries, scientists have wondered how they moved the rocks.

Stonehenge was built in southern England about 5,000 years ago. Scientists think the builders dragged some of the rocks on wooden sleds from 150 miles away.

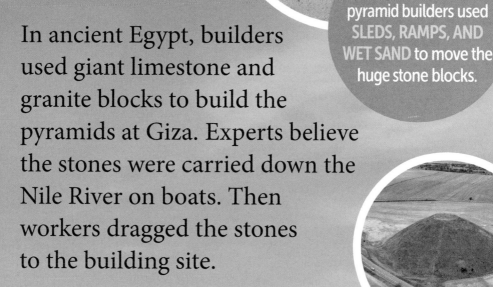

In ancient Egypt, builders used giant limestone and granite blocks to build the pyramids at Giza. Experts believe the stones were carried down the Nile River on boats. Then workers dragged the stones to the building site.

The Egyptian pyramid builders used SLEDS, RAMPS, AND WET SAND to move the huge stone blocks.

England's Silbury Hill is the world's largest ancient mound, a grassy hill built by humans. It's 130 FEET HIGH and made of 500,000 TONS OF CHALK.

Nature also creates incredible formations from rocks and minerals. Sometimes wind and water erode, or wear away, parts of rocks. This erosion can create amazing shapes. For example, Australia's Wave Rock has been carved by more than 100 million years of erosion.

Natural rock formations don't last forever. In 2017, the Azure Window stone arch on the island of Malta COLLAPSED DURING A STORM.

In Turkey, the formations known as "fairy chimneys" look like a magical kingdom. But these spires weren't created by fairies. Instead, wind and rain eroded a top layer of rock formed from volcanic ash. As it eroded, the basalt rock underneath was left standing.

In the Crystal Cave in Ohio, U.S.A., the mineral celestine builds up to form crystals up to A FOOT AND A HALF LONG!

Legend says that the giant rock formations of Ha Long Bay in Vietnam were CREATED BY DRAGONS to keep out invaders.

CHAPTER 2

A Rocky Planet

It can take a single rock MILLIONS OF YEARS to move through the rock cycle.

The oldest known rocks on Earth are 4.28 BILLION YEARS OLD—almost as old as Earth itself!

Rocks on Earth are always changing. They slowly change from one type of rock to another in the rock cycle.

The cycle starts when rocks are squeezed and heated in Earth's crust. The movement of Earth's tectonic (tek-TON-ik) plates forces the rocks up to the surface. There, rocks are broken down by wind, water, and ice. These rocky bits are often swept out to the ocean, where they settle in layers and harden into new rocks. The tectonic plates push the layered rocks back into Earth's crust. Then the cycle starts over!

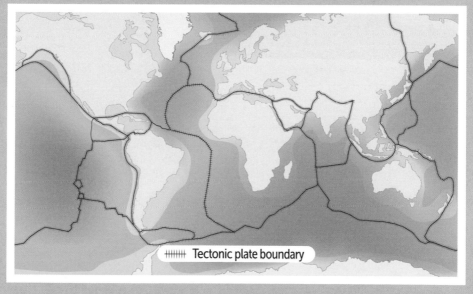

╫╫╫ Tectonic plate boundary

⬆ Earth's crust is made up of 15 to 20 massive rock plates. These tectonic plates fit around Earth like a cracked shell and are constantly shifting.

There are three main types of rock: igneous (IG-nee-us), sedimentary (SED-uh-MEN-ter-ee), and metamorphic (met-uh-MOR-fik). Igneous rock starts out as magma. Sometimes magma erupts through a volcano. When it quickly cools, it becomes rocks like basalt, obsidian, pumice, and rhyolite (RYE-uh-lite).

obsidian

When erupting, some lava can be as hot as 2200 DEGREES FAHRENHEIT.

When lava flows through a forest, a thin layer sometimes cools around burning tree trunks, making HOLLOW MOLDS called LAVA TREES.

Sometimes hot magma doesn't erupt. Instead, it cools slowly and hardens under Earth's crust, creating rocks such as granite, gabbro, or diabase.

gabbro

The **WHITE CLIFFS** of Dover, England, are **MADE OF CHALK**, a type of limestone made up of the tiny bodies of plankton from ancient oceans.

On Earth's surface, a rock doesn't just stay in one piece! Rocks are constantly being churned into small bits by wind and water. Sometimes they get stuck together in layers, which build up over time. This is how sedimentary rocks are formed.

In the Moenkopi Formation in Zion National Park, in Utah, U.S.A., you can see the layers of rock that piled up in an ancient sea. The fossilized footprints of reptiles and even the ripple marks of water are preserved in the rock.

Fossils are almost always found in SEDIMENTARY ROCK.

When rocks sink deep into Earth's crust, high temperatures and pressure heat and squeeze the rocks, which changes them. They become metamorphic rocks.

Some rocks are mashed by Earth's heating and squeezing, so that they have visible FOLDS, SMEARS, OR SWIRLS.

folded limestone

In this way, other types of rock can be "cooked" into metamorphic rock. Limestone, which is a sedimentary rock, can be cooked into marble, a metamorphic rock. Igneous granite can be heated and squeezed into metamorphic gneiss (NISE).

In the 1840s, the Original Monstre Rock Band performed in England on a XYLOPHONE-LIKE INSTRUMENT made from hornfels, a metamorphic rock.

Mineral crystals can be AS LARGE AS A TREE TRUNK or so tiny they are visible only through a microscope.

Minerals

When minerals form, they grow in specific shapes that repeat in patterns. These patterns form an orderly structure called a crystal.

Some minerals are formed from hot magma. As the magma cools, the crystals grow. Other minerals are made when chemicals from nature break down in water. As the water evaporates, or dries up, it leaves behind a solid mineral. Sometimes, a mineral is squeezed or heated by Earth. New ingredients may get mixed in, and a brand-new mineral forms.

Geodes form when BUBBLES IN LAVA cool, harden, and fill with fluid. As the fluid dries up, LARGE MINERAL CRYSTALS grow inside.

pyrite

dolomite

malachite

Minerals form in a variety of crystal patterns. Different patterns result in different-looking minerals. Pyrite grows in perfect cubes. Dolomite can look like grains of rice stuck together. Malachite (MAL-uh-kite) can grow in clumps of green bubbles like peas.

Minerals can be red, gold, bright green, deep purple, blue, yellow, or white. Certain minerals glow different colors under ultraviolet lights.

The yellow, red, or brown mineral willemite can GLOW BRIGHT GREEN under ultraviolet light.

Selenite gypsum (SEL-uh-nite JIP-sum) sometimes forms in THE SHAPE OF ROSES.

Gems are minerals that have been cut and polished. When minerals are first mined, they look very different from the gems in jewelry. They look more like regular stones.

a rock that contains spessartine crystals (left) and a spessartine gem (right)

raw aquamarine (left) and an aquamarine gem (right)

FRESHLY MINED RUBIES look like hunks of pinkish stone.

Diamonds are so hard, they are used to CUT THROUGH OTHER MINERALS AND ROCKS, including gems.

To make a gem, a mineral is first cut using a special saw. Then it is shaped with a metal wheel called a grindstone. After that, the mineral is sanded, then polished until it sparkles.

Put to the Test

People who study rocks and minerals are called geologists. They use many tools and tests in their work.

The Mohs (MOZE) hardness scale helps geologists rank minerals from softest to hardest. By scratching one mineral with another, geologists can see which is harder. This helps identify the mineral.

Calcite is about as hard as YOUR FINGERNAIL.

⬆ In streak tests, the color of the powder is not always the same as the apparent color of the mineral. Here, the black mineral, hematite, leaves a reddish-brown streak.

Geologists might also use a streak test to identify a mineral. In this test, they scratch a mineral across a surface and look at the color of the powder left behind. Each mineral has its own streak color.

talc

Talc is very soft—level 1 on the Mohs scale. DIAMOND is level 10—THE HARDEST.

To figure out how old a rock or mineral is, geologists can look at rock layers. They use clues in the layers, such as certain kinds of common fossils, to tell how old different layers are.

Rock layers are also called **STRATA.**

Radiometric dating is another way of figuring out a rock's age. Rocks have certain elements that break down over set periods of time. Geologists use these time periods like a clock. By studying how much these elements have broken down, geologists can tell how long a rock has existed.

One fossil used to date rock layers is a **TYPE OF CLAM** that lived all over the world 34 to 56 million years ago.

CHAPTER 3

Digging In

a fossilized shell

In Idaho, U.S.A., researchers found a 17-MILLION-YEAR-OLD FOSSIL of a magnolia leaf.

Did you know that fossils are rocks? Many fossils form when animals die and their bodies are covered by layers of sedimentary rock. Water rich in minerals seeps into these animal bones or shells as they decay. The minerals slowly form crystals in the exact shape of the bones or shells.

In 2014, construction workers in Seattle, Washington, U.S.A., found a FOSSILIZED MAMMOTH TUSK while digging a foundation!

Plants can be fossilized, too. When minerals soak into wood, they harden as the wood decays and look just like the wood did!

Fossils can form from nonliving things, too. Scientists have found the fossilized tracks of ANCIENT WORMS.

Some minerals are so rare that the ENTIRE PLANET'S SUPPLY of them could FIT ON A QUARTER.

Common minerals, like quartz, can be found almost anywhere. But other minerals exist only in certain places, under certain conditions. The mineral fingerite forms in only one place on Earth—near the top of the Izalco volcano in El Salvador. And it washes away every time it rains!

Scientists can **GROW DIAMONDS** and other minerals in labs. Lab-grown diamonds can be produced in as little as three months.

Only **ONE TINY CRYSTAL** of the mineral ichnusaite (ik-NOO-say-ite) has ever been found.

The pink mineral cobaltomenite (KOH-bawl-TAH-men-ite) is found in four places on Earth: Argentina, Bolivia, the Democratic Republic of the Congo, and Utah. If you gathered all of it together, it would fit inside a small drinking glass.

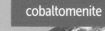

cobaltomenite

In ancient Rome, soldiers were sometimes PAID IN SALT.

Throughout history, some minerals were so valuable, they were used as money. Metallic minerals were often made into coins. As far back as 1000 B.C., people in China were making some of the first known coins out of copper.

Five hundred years later, around 500 B.C., people in what is now Turkey mined chunks of silver, stamped them into round shapes, and used them as money.

You can buy a bar of gold about THE WEIGHT OF A JELLY BEAN for around $50.

The word "salary," meaning the money a person is paid for work, comes from THE LATIN ROOT WORD SAL, which MEANS "SALT"!

Bring the Bling

Diamonds are worth about $3,250 per carat, but the gem painite (PANE-ite) is worth $50,000 PER CARAT!

painite

A gem's weight is measured in carats. One carat weighs about as much as an average raindrop. Often, the rarer a gem is, the higher its value. Painite, for instance, is one of the most valuable gems on Earth and is mined only in Myanmar.

Rubies that are found only in Myanmar have very little iron, so their "PIGEON-BLOOD" COLOR is unusually bright.

People also value gems that have an especially beautiful or unusual appearance. For example, a single ammolite gem can have all the colors of the rainbow.

One of the rarest kinds of sapphires in the world isn't blue—IT'S PINK-ORANGE and found in Sri Lanka, Madagascar, and Tanzania.

Sometimes jewels are so special that legends build up around them. The Koh-i-Noor diamond was mined in India in 1304. People said it was cursed and that it brought death and bad luck to each male ruler who owned it. The Delhi Purple Sapphire (which is actually an amethyst) was thought to be so cursed with bad luck that its owner locked it in seven boxes and surrounded it with good luck charms.

▲ The Koh-i-Noor diamond was once set as the center jewel of an armband. This replica is on display in a museum.

When robbers tried to steal the famous yellow Sancy diamond, the messenger carrying it SWALLOWED THE GEM to keep it safe.

King Louis XVI and Marie Antoinette owned the Hope diamond and were later killed, likely giving rise to the legend that THE DIAMOND IS CURSED.

For thousands of years, people have admired the natural beauty of minerals. That's one reason gold, silver, and gems are used in jewelry. Wealthy ancient Egyptians often wore gold collars decorated with glass and obsidian. Many Native American tribes set turquoise gems in belts and rings.

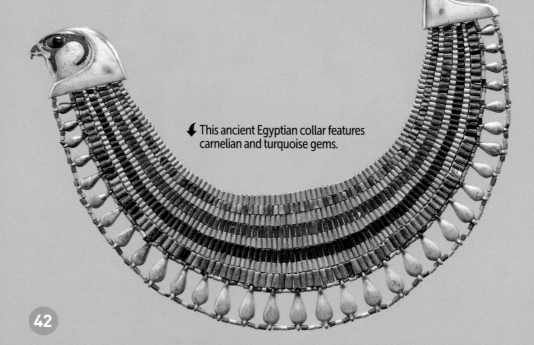

◀ This ancient Egyptian collar features carnelian and turquoise gems.

Ancient Greeks often **CROWNED THEIR DEAD WITH FUNERAL WREATHS** made from thin sheets of hammered gold.

Fashion designer Zac Posen once created an evening gown covered in more than **10,000 SEQUINS MADE OF GOLD.** It was worth $1.5 million!

Whether glittering in a necklace or crowning the summit of a mountain, rocks and minerals bring the bling to our world!

1

The Washington Monument in Washington, D.C., U.S.A., is built from three different kinds of marble.

2

In ancient China, royalty were sometimes buried in suits made entirely of jade plates sewn together with wires.

3

The mineral topaz can range in color from light blue to pink to yellow to colorless!

4

It can take millions of years for magma under the ground to cool down enough to turn solid.

5

Some fossilized dragonflies have a wingspan of two and a half feet—about as long as a person's step.

6

A person who cuts and polishes gems is called a lapidary.

7

According to Maori legend, the round boulders on New Zealand beaches are ancient scraps of gourds and baskets washed ashore from a canoe.

8

Coal is made up of decayed ancient plants.

9

Fossilized wood can be preserved so perfectly that even insect holes may still be visible millions of years later.

10

The mineral bornite is also called "peacock ore" because it turns brilliant blue, purple, and green colors when exposed to air.

25 MORE FACTS ABOUT ROCKS AND

11

The colors in fireworks come from minerals: Barium makes green, and copper makes blue.

12

Teeth and bones are white because they are made mostly of the mineral apatite, which contains calcium.

13

The mineral kaolinite (KAY-uh-luh-nite) is used as an ingredient in some medicines *and* in making kitty litter (but don't eat kitty litter!).

14

Some minerals, such as calcite, can glow in the dark.

15

The ocean floor is made up of basalt—hardened volcanic lava.

16

In parts of the Great Wall of China, the bricks were cemented together with a mixture of heated limestone, water, and sticky rice.

17

Ancient Romans imagined that diamonds were splinters that had fallen off stars.

18

The ancient Egyptians used powdered gypsum, a mineral, to make white paint for tomb paintings.

19

The Cullinan I is the biggest cut diamond in the world. It weighs about as much as a can of soup.

20

Obsidian can be cut into blades that are even sharper than modern surgical scalpels.

21

The oldest fossils ever found were located in 3.5-billion-year-old rock in the Apex Chert formation in Western Australia.

22

Copper kills bacteria and fungi on contact. That's why many hospitals use it on surfaces like bed rails.

23

At the Crater of Diamonds State Park in Arkansas, U.S.A., you can search for precious minerals—and keep everything you find!

24

Chalk, a type of limestone, is used as an ingredient in makeup.

25

A variety of the mineral hackmanite turns from purple-pink to white when exposed to sunlight. In darkness, it turns purple-pink again.

GEMS

Rocks and Gems Facts Roundup

ROCK ON!
You've dug through all the sparkling specifics about rocks, minerals, and gems. Did you catch all 100 facts?

1. A piece of rock has to be at least .16 inch wide to be called a pebble. 2. Cobalt got its name from the German word *kobold*, a mischievous goblin said to live in mines. 3. Archaeologists in Texas found a 1,500-year-old piece of fossilized human poop that contained the bones of a rattlesnake. 4. Powdered aluminum is a main ingredient in rocket fuel. 5. When sand is mixed with other ingredients and heated above 2400 degrees Fahrenheit, it melts and turns into glass. 6. Some people scrub their skin with the volcanic rock pumice. 7. British settlers in Ontario, Canada, called one rock "puddingstone" because it reminded them of a dessert. 8. Diamond is the hardest natural material on the planet. 9. Soapstone gets its name from its greasy, or "soapy," feel. 10. Cinnabar is called "dragon's blood" because of its deep red color. 11. When lightning strikes sand, the sand sometimes melts, forming hollow, glass-lined tubes. 12. Oil can be extracted from the rock shale. 13. At the 2019 Oscars, Lady Gaga wore a $30 million yellow diamond necklace. 14. The gold in Fort Knox was transported there on a heavily guarded nine-car train. 15. Stardust containing minerals rich in iron and nickel regularly falls to Earth. 16. Freezing weather can break apart rocks. 17. Prehistoric peoples made axe-heads from jade to use at special ceremonies. 18. Hildegard von Bingen believed that licking a sapphire would make a person smarter. 19. An average American uses about 40,000 pounds of minerals every year. 20. Clay is soft because it's made up of tiny bits of rock clinging together! 21. Ancient Egyptians said tourmaline passed through a rainbow, giving it its many colors. 22. Volcanic rocks are so hard that ancient Romans built roads with them. 23. Mineral crystals can look like ferns or bicycle spokes. 24. Mica gives some makeup its shimmer. 25. Ulexite is called "television stone"—images placed underneath it seem to appear on its surface. 26. Rocks and minerals can be found even in space! 27. Our planet contains over 5,000 known minerals. 28. Rocks from other planets slam into Earth in the form of meteorites. 29. People started using rocks to make tools two million years ago. 30. The Navajo people have been mining and using turquoise since at least 200 B.C. 31. Silver is used in medical bandages to prevent infection. 32. Some of the giant rocks of Stonehenge weigh about 50 tons each! 33. Egyptian pyramid builders used sleds, ramps, and wet sand to move the huge stone blocks. 34. England's Silbury Hill is the world's largest ancient mound. It's 130 feet high and made of 500,000 tons of chalk. 35. In 2017, the Azure Window stone arch collapsed during a storm. 36. Legend says the giant rock formations of Ha Long Bay were created by dragons to keep out invaders. 37. In the Crystal Cave in Ohio, celestine forms crystals up to a foot and a half long! 38. The oldest known rocks on Earth are 4.28 billion years old! 39. It can take a single rock millions of years to move through the rock cycle. 40. Fiery-hot liquid rock exists miles below our feet. 41. Lava can be as hot as 2200 degrees Fahrenheit. 42. When lava flows through a forest, a thin layer sometimes cools

around burning tree trunks, making hollow molds called lava trees. 43. The White Cliffs of Dover are made of chalk. 44. Fossils are almost always found in sedimentary rock. 45. Rocks can be squeezed or stretched like taffy. 46. Some rocks have visible folds, smears, or swirls. 47. The Original Monstre Rock Band performed on a xylophone-like instrument made from hornfels. 48. Mineral crystals can be as large as a tree trunk or so tiny they are visible only through a microscope. 49. Geodes form when bubbles in lava cool and harden, and large mineral crystals grow inside. 50. Some minerals look like soft balls of cotton—but they're made of crystal needles! 51. Yellow, red, or brown willemite can glow bright green under ultraviolet light. 52. Selenite gypsum sometimes forms in the shape of roses. 53. Freshly mined rubies look like hunks of pinkish stone. 54. Diamonds are used to cut through other minerals and rocks. 55. Calcite is about as hard as your fingernail. 56. Talc is very soft—level 1 on the Mohs scale. Diamond is level 10—the hardest. 57. Rock layers are also called strata. 58. One fossil used to date rock layers is a clam that lived 34 to 56 million years ago. 59. In Idaho, researchers found a 17-million-year-old fossil of a magnolia leaf. 60. In 2014, construction workers found a fossilized mammoth tusk while digging a foundation! 61. Scientists have found the fossilized tracks of ancient worms. 62. Some minerals are so rare that the entire planet's supply of them could fit on a quarter. 63. Scientists can grow diamonds and other minerals in labs. 64. Only one tiny crystal of ichnusaite has ever been found. 65. In ancient Rome, soldiers were sometimes paid in salt. 66. You can buy a bar of gold about the weight of a jelly bean for around $50. 67. The word "salary" comes from the Latin root word *sal*, which means "salt"! 68. Painite is worth $50,000 per carat! 69. One of the rarest kinds of sapphires in the world isn't blue—it's pink-orange. 70. Rubies found only in Myanmar have very little iron, so their color is unusually bright. 71. When robbers tried to steal the Sancy diamond, the messenger swallowed the gem to keep it safe. 72. King Louis XVI and Marie Antoinette owned the Hope diamond and were later killed, likely giving rise to the legend that the diamond is cursed. 73. As far back as 2500 B.C., Sumerian people wore gold and lapis lazuli earrings and headdresses. 74. Ancient Greeks often crowned their dead with funeral wreaths made from thin sheets of hammered gold. 75. Zac Posen once created an evening gown covered in more than 10,000 gold sequins. It was worth $1.5 million! 76. The Washington Monument is built from three different kinds of marble. 77. In ancient China, royalty were sometimes buried in suits made entirely of jade plates. 78. Topaz can range in color from light blue to pink to yellow to colorless! 79. It can take millions of years for magma under the ground to cool down enough to turn solid. 80. Some fossilized dragonflies have a wingspan of two and a half feet. 81. A person who cuts and polishes gems is called a lapidary. 82. According to Maori legend, the round boulders on New Zealand beaches are ancient scraps of gourds and baskets washed ashore from a canoe. 83. Coal is made up of decayed ancient plants. 84. Fossilized wood can be preserved so perfectly that even insect holes may still be visible millions of years later. 85. Bornite turns brilliant blue, purple, and green colors when exposed to air. 86. The colors in fireworks come from minerals: Barium makes green, and copper makes blue. 87. Teeth and bones are white because they are made mostly of apatite, which contains calcium. 88. Kaolinite is used as an ingredient in some medicines *and* in making kitty litter. 89. Some minerals, such as calcite, can glow in the dark. 90. The ocean floor is made up of basalt. 91. In parts of the Great Wall of China, the bricks were cemented together with a mixture of heated limestone, water, and sticky rice. 92. Ancient Romans imagined that diamonds were splinters that had fallen off stars. 93. The ancient Egyptians used powdered gypsum to make white paint for tomb paintings. 94. The Cullinan I is the biggest cut diamond in the world. 95. Obsidian can be cut into blades that are even sharper than modern surgical scalpels. 96. The oldest fossils ever found were located in 3.5-billion-year-old rock in Western Australia. 97. Copper kills bacteria and fungi on contact. 98. At the Crater of Diamonds State Park, you can search for precious minerals—and keep everything you find! 99. Chalk is used as an ingredient in makeup. 100. A variety of hackmanite turns from purple-pink to white when exposed to sunlight. In darkness, it turns purple-pink again.

INDEX

Boldface indicates illustrations.